Science of Electricity

Volume 3

Hydropower Explained Simply

by Mark Fennell
© 2012

This book is part of the
Energy Technologies Explained Simply™ Series

Other Books in the Energy Technology Series

About the Book

Hydroelectric power is one of the oldest methods of creating electricity and it is still one of the best. There are virtually no environmental problems, there are only a few mechanical parts, and yet we can get large amounts of electricity from it. However, in order to make hydropower effective there are some practical factors to consider.

This book discusses the most important concepts of hydropower. The book begins by explaining the basic concepts of hydropower, and showing the reader how hydropower works.

The second chapter discusses dams and rivers. Related topics include reservoirs, spillways, and removal of debris. Dams provide the necessary storage and height, while the rivers provide the flow. Dams are used to harness the power of water, yet must be designed, managed, and maintained properly.

The third chapter explains how to calculate the amount of power from a hydropower system, and offers essential hydropower calculations for any situation.

The fourth chapter provides an overview of micro-hydro systems, with some practical tips for those readers who are considering installing such a system. Micro-hydro systems are growing in popularity world-wide, serving individual homes and small communities. This chapter provides an overview of factors to consider when installing your micro-hydro system.

The final chapter discusses turbines in detail. Using a more efficient turbine will create more electricity. Whether you are building a micro-hydro system or a megawatt power plant, it is important to choose the proper turbine for your particular situation. In addition, some of these turbines are used in other types of power systems, such as coal or natural gas.

About the Energy Technology Series

<u>Purpose of this Series</u>

The books in the *Energy Technologies* series are designed to educate citizens, students, and legislators on all aspects of energy technologies. The first books in the series focus on electrical power.

The books discuss many energy technologies, including: generators, turbines, power plants, power lines, and grids. The technologies for each type of power source (hydro, wind, solar, coal, nuclear, and natural gas) are discussed in detail. The books also discuss efficiency, safety, reliability, and health concerns for each energy technology.

The ultimate goal of the series is to enable the people to make informed decisions on practical energy questions. The secondary goal is to serve as introductory guides for students embarking on careers with energy technologies.

Taken altogether, the books in the series answer any question you are likely to have, such as:

- How can we increase the efficiency of solar cells?
- How do I select the size my solar array?
- What do I need to know when installing a wind turbine?
- How effective are the clean coal technologies?
- How can we prevent grid failures?
- Do power lines cause cancer?
- and many other energy technology questions...

<u>Science of Electricity in Perspective</u>

The subject of electrical power is of great importance to our communities, but is rarely taught. At best, an informed citizen knows only a few pieces. At worst, as it is for a great number of citizens, electricity is magic and myths are believed as scientific truth. It does not have to be that way. Any citizen, regardless of background, can know the technologies behind all aspects of electricity.

The books in this series solve that problem. These books educate the general public in all aspects of electrical power. Any person, regardless of background, can easily find the answer to his energy question in one of these books.

Specific Goals

There are numerous technologies described in these books. Yet for each technology I sought out the answers to the following questions:

1. How does the technology work?
2. What are the advantages and disadvantages?
3. What is the efficiency? How can the efficiency be improved?
4. What is the environmental impact? How can it be improved?
5. What are the safety hazards, and how can they be reduced?
6. What are the most important practical tips?
7. What facts comprise the most important data?

Technical Discussions Explained Simply

The books in the series must necessarily be technical to some degree. Electricity is a practical technology, and therefore we must understand the technical aspects if we want to make wise decisions. Yet the discussions in this book are always aimed at the citizen or policy maker.

The books in this series explain the principles of electricity as simply as possible, using ordinary English (no engineering jargon), and highlighting the most important points of each technology. Main concepts and facts are emphasized with the use of lists, tables, diagrams, and summaries.

I do not expect any reader to have a background in science, yet I offer enough facts and details so that the reader can have an accurate understanding of all related technologies. I provide enough technical details and enough data for the reader to make informed decisions.

Conclusion

For all the reasons above, I offer this series of books. My goal is to inform you on the basic concepts of all the technologies and all of the issues related to electricity so that you can make realistic decisions.

Remember that there are no perfect solutions, there are only choices. I hope that this series of books will assist you in making those choices for your community.

Mark Fennell

Accuracy and Technical Depth

Objectivity

I have tried my best to be as objective as possible. Whereas many other authors of energy books have an agenda, I have no desire to promote one industry over another. I have no desire to promote one technical solution over another. In this endeavor, I have tried to be an objective scientist.

Accuracy of Data and Summaries

I never relied solely on the conclusions of other researchers. Instead, I performed many other tasks to ensure that all conclusions were accurate. I examined primary data whenever possible. I have read the fine print on how research was obtained.

I have also checked the accuracy of the conclusions written by other researchers, most commonly by finding at least three distinct sources for each fact. In addition, I performed my own calculations numerous times to prove (or disprove) conclusions and final values in other reports. It is only after such rigorous investigations that I created data tables and wrote summaries for these books.

Limited Mathematics

The books must also use math from time to time. For example, efficiency is a statement of a specific amount, and therefore the discussion of efficiency requires the use of equations. Other issues such as power loss, health hazards, environmental concerns, and quality control are also statements of amounts and also require calculations.

Therefore some equations are necessary to know, even for the non-scientist. I also provide examples of calculations so that readers can become more comfortable with using the equation themselves.

However, I want to emphasize that I focus on concepts not on the mathematics. I provide equations only when it is necessary for the citizen or student to be familiar with these equations.

M.F.

Table of Contents

3.1
Basics of Hydroelectric Power

Introduction

Hydroelectric power is one of the oldest methods of creating electricity and it is still one of the best. There are virtually no environmental problems, there are only a few mechanical parts, and yet we can get megawatts of electricity from it.

Brief Description of Hydroelectric Power (Fig. 3.1)

Hydroelectric power takes the basic principles of a waterfall and harnesses those principles into electrical power:

1. A reservoir stores water, at a location much higher than the river below. The reservoir height is called the head.
2. Water flows down through a channel, called a penstock.
3. The flowing water pushes the blades of the turbine.
4. The rotation of the turbine then turns the magnets in the electric generators, thus producing electricity.

A Closer Look at the Process

We build a dam in order to create a difference in height between the reservoir and the turbines. This difference in height allows the water to fall, much like a natural waterfall. The falling water transforms the water's potential energy into kinetic energy.

The water is then channeled through the dam (not over it) in a pipe called the "penstock." Beyond the penstock is a door to the turbine, called the "sluice gate." When the sluice gate is open, the water flows to the turbine. The water pushes the turbine blades, causing the turbine to rotate. The rotating turbine operates the electrical generator, which creates the electrical power. After the water leaves the turbine, the water is sent to the river to follow its normal course.

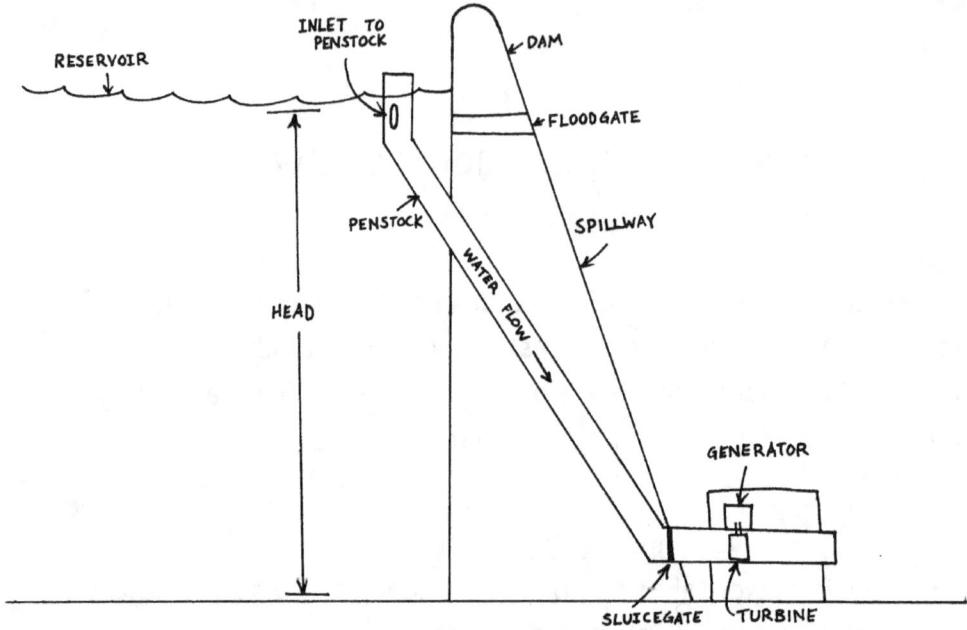

Figure 3.1   Basic Hydropower Plant

Key Factors in the Power from Hydroelectric Plants

The amount of power from a hydroelectric plant depends on the rate of flow of the water, and the distance the water falls (the "head"). The faster the flow of water, the more power we can get. The higher the dam, the more power we can get. "Rate" and "Head" are measured as follows:

1. The rate of the water: The flow rate is measured at the point where the water goes through the sluice gates and into the turbines. The rate is usually measured as a volume of water per time. Common units of flow include liters per second, and gallons per minute.

2. The distance which the water falls (the "head"): The distance which the water falls is called the "head." It is usually measured in meters. The height of the head is essentially the height of the dam. However, do note that the "rated head" (the value actually used for power calculations) is usually a bit less than the dam height. There are several reasons for this, including: specific design of the hydropower system, varying seasonal levels of the reservoir, and friction in the penstock.

Specific calculations for power from a hydroelectric plant will be discussed in another chapter. However, at this point we can state the basic formula for power: Power = head x flow x efficiency x conversion factor.

Relationship Between Flow of Water and Height of Dam

Note that because of the relationship between flow of water and height of the dam, you can work with either the height of the dam or the flow of the water to get your results. You can either have 1) a tall dam with a slow river, or 2) a short dam with a rushing river, and get the same results. For example, these two dams get the same amount of power:

a) Slow river, tall dam: flow of 60 gallons per minute, dam 700 feet high; power = 4.6 kilowatts

b) Fast river, short dam: flow of 6,000 gallons per minute, dam 7 feet high; power = 4.6 kilowatts

Thus, you can get a significant quantity of energy from a small dam, as long as you have sufficient water flow in your river. However, for generating large amounts of power it most effective to build the tallest dams possible.

Chapter Summary

1. In hydroelectric power generation, a dam is built in order to create height and a reservoir. The water flowing from the dam pushes the turbine blades. The rotation of the turbine drives the generator.

2. The channels that the water flows through are called penstocks, and the openings to the turbines are called sluice gates.

3. The amount of power from a hydroelectric plant depends on the rate of flow of the water and the distance that the water falls.

4. The height of the water is called the head. For approximate calculations, the head is essentially the height of the dam.

5. The basic formula for power from a hydropower plant is:
   Power = head x flow x efficiency x conversion factor

3.2
Dams, Reservoirs, and Rivers

Introduction

The basic parts of the hydropower system include: the dam, the reservoir, the sluice gate, the spillway, the penstock, the trash rack, and the river. When you first look at a dam you will see not only the dam and the reservoir, but also the spillway and the power plants.

The center part of the dam is known as the spillway. The spillway exists to allow excess water to flow to the river below. You will notice that the actual power plants are usually located on the left and right of the dam. It is on the sides of the dam, not the center, where the water flows to the turbines. It is on the sides of the dam where the electrical power is generated.

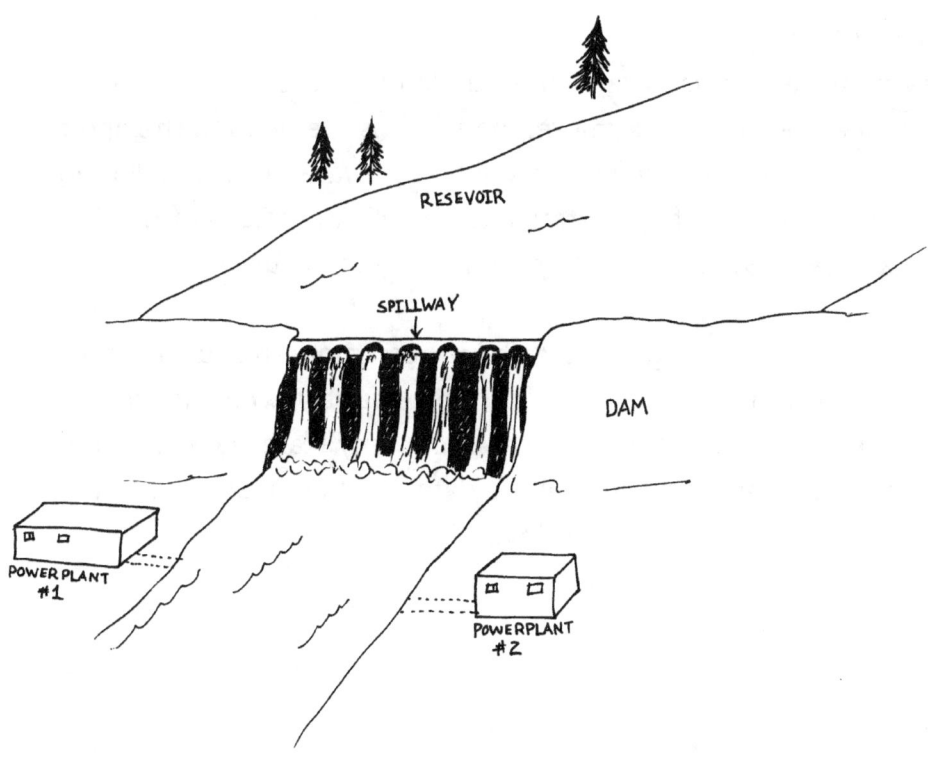

Figure 3.2 Dam, Power Plant, and Spillway

Dams

Introduction

The main purpose of a dam is to provide height. The higher the dam, the greater power we can get. Dams must also resist a great amount of pressure from the water stored in the reservoir. The factors in the dam design include: dam type, material, and spillway mechanisms.

Dam Types

Dam types include: arch, buttress, earth, and gravity. The most common dam type for hydro-power is the "gravity" dam, where the dam is thicker at the bottom. The tallest of dams (greater than 300 feet) are usually arch type dams. The material for a hydro-power dam is usually concrete. However, within the concrete there can exist several types of materials to reinforce the dam. These materials include metals, plastics, and engineered composites.

Inspection of Dams

Dams must be inspected regularly. All dams weaken with age and will eventually need repair. Furthermore, many dams are nearing the end of their designed lifetimes. Engineers are already aware of this, including the dam safety officials of each state and the US Army Corp of Engineers. These authorities inspect and repair dams daily to ensure that our dams are safe.

Most of the nation's largest dams have been modernized. However, there are approximately 90,000 dams in this country, which is an enormous number of dams to inspect. Also, many states lack the funds to properly inspect and maintain all the dams in their area. This is a situation which needs greater scrutiny.

Controlling Water Flow: Sluice Gate and Spillways

Introduction

Building a dam is essentially building an artificial waterfall. In fact, a dam is really two artificial waterfalls. The first waterfall provides the energy for the turbines. The second waterfall allows excess water to flow past the dam into the river below. It is important that we control the flow of both of these artificial waterfalls (reasons explained below). It is the jobs of the sluice gate and spillways to control that flow.

Controlling Water Flow: Sluice Gate (figure 3.1)

The sluice gate controls the flow of water that reaches the turbine. The "sluice" is the final segment of the penstock, the piece that leads directly to the turbine. The "sluice gate" is essentially a door placed just before the turbine.

If we open the sluice gate wider, then the rate of flow to the turbine will be faster. This results in greater power. If we close the sluice gate a bit, then the rate of flow to the turbine will be slower. This results in less power. We can also close the sluice gates altogether. Most large hydropower plants have several turbines, and hence several sluice gates. If the power is not needed, then we can simply close off access to some of the turbines.

Controlling Water Flow: Spillways and Flood Control

If we just dammed up the reservoir then the water from heavy rain would spill over the sides and flood the area below. Instead, we want to control the flow of water, which would then control the flood. We do this by allowing more water to flow through the dam, by opening the spillways. These spillways allow the extra water to flow through the dam as needed. There are generally three forms of spillways. These can be categorized as: water flowing over the dam, water channeled through the dam, and water channeled around the dam.

The primary form of spillway is literally a small waterfall. The water literally flows over the dam and down the other side. You will notice that the primary spillway takes up most of the length of the dam.

The second form of the spillway consists of openings in the dam, commonly known as "floodgates." Every dam has several channels through the dam, with doors leading to the outside. When more water needs to be let out, the doors are open and water comes through the dam. The third form of spillway is simply a meandering pipe which reroutes the excess water. This pipe is generally placed on the side of the hill, leading from the reservoir around the dam, and finally down to the river or to an overflow tank.

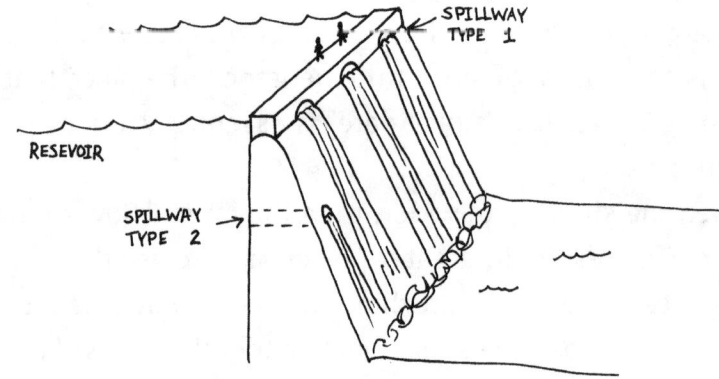

Figure 3.3a Types of Spillways 1&2

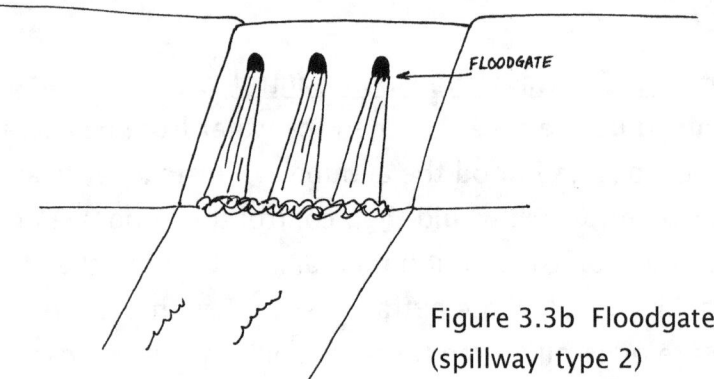

Figure 3.3b Floodgate
(spillway type 2)

Penstocks and Trash Racks (figures 3.1, 3.4)

The penstock is part of the dam system. The penstock is essentially a pipe which channels the water from the reservoir to the turbine. In megawatt hydropower plants the diameter of the penstock can be quite large, often 10 to 40 feet in diameter. The large size of the penstock allows greater volume of water, and thus greater flow, which results in the production of more power. Larger penstocks also have the advantage of lower friction, which also gives us greater power.

Many types of natural debris float down a river. This includes rocks, sticks, leaves, plants, and large branches. We do not want branches or moss to clog up our penstock. Nor do we want rocks to hit the turbine and cause damage to the turbine. Therefore, all reservoirs must have a system of trash racks, screens, and other methods of keeping the debris from entering the penstock.

The first line of defense is usually a trash rack. The trash rack is a relatively simple design, often just a framework of poles. This trash rack stops the larger items such as branches and plant material. The trash rack is usually attached to a pulley, which allows workers to pull the debris closer to shore. These areas can be then be cleaned out periodically. In large power plants this system is usually automated.

The second defense is a small screen around the penstock. This screen has very small holes. As such, it acts like a filter: water will go in, but rocks will stay out. This screen has the additional benefit of preventing any fish from being flung down into the turbine.

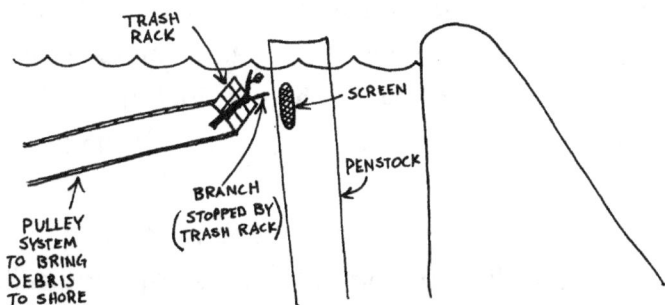

Figure 3.4   Penstock, Trash rack, and Screen

Reservoirs

Introduction

The main purpose of a reservoir is to store water for future use. We collect water during the periods of rain, then we can use the water throughout the year to create electricity. The reservoir can also be used to supplement water needs for agriculture.

There is more to creating a reservoir than just the dam. If we just dammed up a river the water would flood the adjacent areas. Therefore, the reservoir must be built deep and wide. It is for this reason that valleys are often turned into reservoirs.

Sedimentation

All rivers naturally bring some soil mixed in with the water. This soil settles in various parts of the river, including the reservoirs. When soils settle in a region, the soils are collectively called "sediment," and the process is called "sedimentation."

We must remember that the primary purpose of the reservoir is to have a large body of water as storage. However, with that storage of water also comes storage of the settling soils. As sediment builds in the reservoir, the reservoir automatically becomes smaller. As the reservoir becomes smaller, we will not have as great a storage capacity for water as we should. If the sedimentation is unchecked, then the sediment will fill the reservoir, leaving little storage space for the water. A second problem is flooding. A reservoir with years of sedimentation is smaller than designed and will flood very quickly during heavy rains.

The USDA has estimated that the storage loss in a reservoir due to sedimentation is .5%–3.5% storage loss per year. The exact value of sedimentation rate depends on several factors, including: the reservoir design, the flow of rivers that feed the reservoir, and the types of soil along the river.

Therefore, we must measure the rate of sedimentation in nearby reservoirs, and dredge sediment from the reservoir periodically. Citizens must pay for this expense in order to make sure that the size of the reservoir is as large as it can be.

Rivers

Rivers provide the flow of water. Choosing which river to use and choosing what location of that river requires a lot of investigation. Remember that the power from water is based on two factors: speed of the river and height of the dam. A faster river is more preferable to a slower river. However, if we can build the dam high enough then a slow river can be suitable.

Before choosing a location, engineers must take flow data throughout an entire year at any given location. Only after the flow has been measured for all times of the year can we make a realistic determination on whether the site is useful or not.

We must always stop the river where we wish to build the dam, and reroute the river until the dam and power plant are completed. In many cases it is preferable to build a new river. This river would be an extension of the main river. The second river, being smaller, makes maintenance of the power plant much easier. This river can then be routed around to go back into the main river. Alternately, this second river can be used as means of irrigation.

Summary

1. The main purpose of a dam is to provide height. The higher the dam, the greater power we can get.

2. Dams must resist a great amount of pressure from the water in the reservoir. There are several methods available to achieve this.

3. Dams must be inspected regularly. However, many states do not have adequate funds for proper inspection and maintenance of their dams.

4. The sluice gate controls the amount of water that reaches the turbine. The rate of flow of the water is measured at the sluice gate.

5. Opening more sluice gates creates more power. When the power is not needed some sluice gates should be closed so that we do not waste water from the reservoir.

6. The spillway is the method for excess water to flow past the dam. There are three forms of spillways: water flowing over the dam, water being channeled through the dam, and water channeled around the dam. The second type of spillway system is often known as floodgates.

7. The penstock is a pipe which channels the water from the reservoir into the turbine. The larger the penstock, the greater power we can get.

8. All reservoirs must have a system of keeping the debris from entering the penstock. Methods include the use of trash racks and screens.

9. The main purpose of a reservoir is to store water for future use.

10. We must dredge the sediment from the reservoir periodically in order to make sure that the size of the reservoir is as large as it can be.

11. We must always stop the river where we wish to build a dam, rerouting that river until the dam is completed.

12. One of the great advantages of water power is that we are not using up any water. The water is used, but then sent right back into the river.

13. In many cases it is preferable to build a new river, which is an extension of the main river. This river can then be routed around to go back into the main river. Alternately, this river can be used as minor means of irrigation.

3.3
Power Calculations for Hydropower

Introduction

The amount of power from a hydropower plant depends on two factors: rate of the water, and distance the water falls. In this chapter we will look at how each factor is measured, as well as how to calculate the power coming from any hydropower plant.

The Rate of the Water

The rate of the water flow is measured as it goes through the sluice gates and into the turbines. The rate is usually measured as a Volume of water per Time. Common units for flow rate include:

    a. Liters per second, abbreviated: "L/s"
    b. Cubic feet per second, abbreviated: "ft^3/s"
    c. Cubic meters per second, abbreviated: "m^3/s"
    d. kilograms per second, abbreviated: "kg/s"
    e. cubic feet per minute, abbreviated: "ft^3/m" or "cfm"
    f. gallons per minute, abbreviated: "gal/min" or "gpm"

Volume and Flow Unit Comparisons

Introduction

Units of measurement for volume and flow vary widely. Different people choose to use different units. In order to better understand the relative size of these units, I have provided two tables. The first table compares all units of volume to the size of the liter. The second table compares all units of flow to the unit of liters per second. Note that exact conversions for all units will be found in the Appendix.

Unit of Volume	Equivalent in Liters
1. Liters (smallest unit of volume)	1.0 Liters
2. Gallon	3.78 Liters
3. ft^3	28.3 Liters
4. m^3 (largest unit of volume)	1,000 Liters

Unit of Flow	Equivalent in L/sec
gallons/min (smallest units)	.063 Liters/sec
ft³/min	.471 Liters/sec
kg/s	1.00 Liters/sec
L/s	1.00 Liters/sec
ft³/s	28.5 Liters/sec
m³/s (largest units; fewer of them)	1,000 Liters/sec

Design Flow and % Flow

Design Flow

The "design flow" is the flow that you can expect from your river for most of the year. The flow rate changes throughout the year, yet we need one value for the technicians to work with. It doesn't do any good to take the highest flow values because these only last a short time. Instead, we graph the flow rates as related to the time of year. From that chart you pick the value of flow rate that occurs most of the year. That value is your design flow.

% flow

You will often read data referring to "% flow." This value is the percentage of the designed flow which you have at a particular moment. For example, if your "design flow" is 100 gallons per minute, then "30% flow" would be 30 gallons per minute.

We can also have a % flow which is greater than 100%. For example if your design flow is 100 gallons per minute, then "104% flow" would be 104 gallons per minute. Remember that our design flow is not the fastest flow, but rather the flow that occurs most often, so a flow greater than 100% design flow is possible. In fact, flows greater than 100% of the design flow typically occur every year. (This occurs during the seasons when the river runs most swiftly.)

The Head (Distance the Water Falls)

The distance which the water falls is called the "head." The head is essentially the height of the dam. However, the "rated head" (the value actually used for power calculations) is usually a bit less than the dam height. There are several reasons for this, including: specific design of the hydropower system, varying seasonal levels of the reservoir, and friction in the penstock.

The "Gross Head" is the simplest measurement: the height of the dam. The "Net Head" is the more accurate head value to be used in calculations. The main factors which create differences between the Net Head and the Gross Head are pipe friction and bends in the pipe. The Net Head is also referred to as the "Rated Head."

The head is usually measured in meters or in feet. The conversions between meters and feet can be found in the appendix.

Calculating Power

Basic Equations

The basic concept of power calculations is relatively simple:
Power=flow x head. This calculation is made more accurate by including the efficiency. Thus: Power = flow x head x efficiency.

Hydropower is typically measured in terms of kilowatts. In order to obtain the final value in kilowatts we must use a "factor." The "factor" takes into account the particular units of flow and the particular units of head in order to come up with a value for power that is in terms of kilowatts. For example, if the flow is measured in Liters/second, and the head is measured in meters, then the factor is .0097.

The final power equation is: # kw = flow x head x efficiency x factor. This is the general form of the equation, which applies to all situations.

Below is a list of power equations. Each equation uses a specific unit of flow, a specific unit of head, and the necessary factor to arrive at power in terms of kilowatts.

gallon/minute and feet

If the flow is measured in gal/min and the head is measured in feet, then: # kw = (# gal/min) x # feet x efficiency x .0001886

ft³/minute and feet

If the flow is measured in ft³/min and the head is measured in feet, then: # kw = (# ft³/min) x # feet x efficiency x .001411

Liters/second and meters

If the flow is measured in L/sec and the head is measured in meters, then: # kw = (# Liters/sec) x # meters x efficiency x .009745

kg/sec and meters

If the flow is measured in kg/sec and the head is measured in meters, then: # kw = (# kg/sec) x # meters x efficiency x .009745

ft³/s and feet

If the flow is measured in ft³/s and the head is measured in feet, then: # kw = (# ft³/s) x # feet x efficiency x .08467

m³/s and meters

If the flow is measured in m³/s and the head is measured in meters, then: # kw = (# m³/sec) x # meters x efficiency x 9.745

Summary

1. The amount of power depends on:
    a. Rate of flow of the water.
    b. Distance the water falls. (The "head")

2. The most common units for flow, in order of size, are as follows:

Unit of Flow	Equivalent in L/sec
gallons/min	.063 Liters/sec
ft³/min	.471 Liters/sec
kg/s	1.00 Liters/sec
L/s	1.00 Liters/sec
ft³/s	28.5 Liters/sec
m³/s	1,000 Liters/sec

3. The "design flow" is the flow that you can expect from your river for most of the year.

4. The "% flow" is the percentage of the design flow which you have at a particular moment.

5. The distance which the water falls is called the "head." The head is essentially the height of the dam.

6. The "net head" or "rated head" is the value actually used for power calculations. The Net Head value is usually a bit less than the dam height, due to design of the hydropower system and friction in the penstock.

7. The amount of power from a hydropower plant can be calculated by:
    • Power = flow x head.

8. The power equation is made more accurate by including the efficiency:
    • Power = flow x head x efficiency

9. To get power calculated in terms of kilowatts we multiply by a factor.
In general: # kilowatts = flow x head x efficiency x factor

3.4
Micro-Hydro

Introduction

Micro-hydro is hydropower on a small scale. Generally, micro-hydro is designed for one family, but sometimes it is done for a small community. The specific wattage in the definition of micro-hydro varies depending on who you talk to, but a typical definition places the maximum output at 100 kilowatts.

Micro-hydro is tedious to install and maintain. However, micro-hydro can be effective in the right locations. The cost of a micro-hydro installation is expensive, but it generally pays for itself in 7-10 years. Micro-hydro systems have been known to last 15 to 30 years before major parts need to be replaced, which gives the owner several years of free electricity.

Measuring for Feasibility and Design

Before you decide to build a micro-hydro at a particular site, you will need to do a feasibility study. It is important that you take some measurements in order to see if the site can realistically provide enough power. Not only will you need some measurements to ensure feasibility, you will also need complete and accurate data in order to design the system. The turbine design and the pipe system will depend entirely on the specific situation at that location. Therefore, the flow rates and the head must be measured very accurately.

If you talk to a reputable micro-hydro installation company, the first thing they will tell you is that each micro-hydro project is different. Many factors must be examined. The flow rates, the head, the terrain, and even the weather history must be taken into account. There is no such thing as a "standard" micro-hydro installation, not even for sites on the same river, because every location is different.

There are many good guides which tell you exactly how to get accurate data. These guides tell you how to accurately measure the flow, and how to survey the terrain.

However, I suggest hiring a company to do the measurements for you. There are many good micro-hydro companies in various regions of the country who will survey the land and measure the flow rates for you. Paying the technicians for their expertise at the beginning of the project is far better than installing a poorly designed hydropower system. Many do-it-yourself micro-hydro users have made mistakes which they regretted later.

Changing the River, Building Dams

Creating a micro-hydro system really is a civil engineering project, just on a smaller scale. First, a portion of the river must be blocked (or rerouted altogether) in order for construction to begin. Then a dam must be built. Even if it is a small dam, the process of building the dam requires moving heavy rocks, delivery of large amounts of lumber, and lots of physical labor. In addition, even the smaller dams require a certain amount of concrete. Always note that the bigger the dam, the greater amount of civil engineering required.

In the end, the river you knew will be transformed. There will be a dam (possibly made of concrete), a reservoir, and a penstock. All of this alters the river from its natural state, just as in the large hydropower plants.

It is a trade-off between size and power. Large power plants require the flooding of valleys and building large concrete dams, yet we can obtain enough power for thousands of citizens. In contrast, a smaller change to the river leaves more of the natural world intact, but the system produces less power.

Note that flooding is a serious problem with micro-hydro systems. Generally, the rivers where micro-hydro systems are placed are not very deep and not very wide. Therefore, flooding is common. Micro-hydro instillation companies will offer suggestions on how to minimize flooding damage in your location. Again, the solutions are specific to the terrain and weather of the location.

Turbine Selection

Selecting the turbine is one of the most important parts of the entire micro-hydro process. Always remember this concept: every micro-hydro turbine is custom built for a specific location. No two are exactly alike. If you are actually going to install a micro-hydro system then you must work closely with a turbine company. Together you will come up with the best turbine for your location.

The exact choice of turbine, the size, the blade design, the number of jets, and choice of other features depends entirely on the flow and the head at your location. Experts have compiled empirical data over the years based on actual turbines in use. This extensive data tells you which turbines work best in which circumstances. (For general principles behind the major types of turbines see the chapter in this unit on Turbines.)

Batteries and Emergency Generators

The micro-hydro system differs from large power plants in that we cannot regulate the flow of water. In the large power plants we can close access to various turbines as needed. With micro-hydro, the water keeps flowing, the turbine keeps spinning, and the electricity is wasted. Therefore, all users of micro-hydro power should have a series of batteries. Any extra power produced by the system will automatically go to these batteries. No energy is wasted. The batteries can then be used when extra power is needed, or when the river flow is low.

Everyone who uses a micro-hydro system must also have an emergency generator. This will be a diesel generator, with enough fuel to last for several days. You never know when a severe storm will come. If your turbine system goes out or becomes clogged during the storm then you don't want to go outside to make repairs. With a generator, you can wait out the storm in safety.

Chapter Summary

1. Micro-hydro is hydropower on a small scale. Generally, micro-hydro is designed for a single family or for a small community.

2. Before you decide to build a micro-hydro at a particular site you must do a feasibility study. In order to design the micro-system at your location, you will need complete and accurate data.

3. Selecting the turbine is one of the most important parts of the entire micro-hydro process. Each micro-hydro turbine is built for a specific location, for a dam with a specific head, and for a river with a specific flow.

4. All users of micro-hydro power should have a series of batteries to store extra power produced, and diesel generators for emergencies.

3.5
Turbines for Hydropower

Introduction

Using a more efficient turbine will create more electricity. Whether you are building a micro-hydro system or a megawatt power plant, it is important to choose the proper turbine for your particular situation. Therefore, it is valuable to understand the basic turbines that are available. In this chapter we will discuss the most common types of turbines used in hydropower systems.

General Points on Selecting Turbines

Any one of the turbines discussed in this chapter may be used in a power generating station. The choice depends on the circumstances. Many of these turbines have good efficiency, usually 80% or more. However, the high efficiencies are stated with the assumption that you use the turbine in the proper circumstances.

Every turbine is custom designed. You can never just buy a "standard" of any type of turbine. Whether the turbine is for a single family home or for a megawatt power plant, the turbine must be custom designed for the particular location. For example, if you wanted a Pelton turbine then you would have to order it. It is true that any Pelton turbine would have the basic features of the Pelton design. However, the exact size would be based on the rated head and the design flow of your particular hydropower plant.

The turbines discussed include:
- Classic Water Wheel (figure 3.5a,b)
- Pelton Wheel (figure 3.6)
- Turgo (figure 3.7)
- Propeller – fixed pitch (figure 3.8)
- Kaplan – also propeller, yet with adjustable blade (fig 3.9)
- Francis (figure 3.10)
- Cross-Flow (Figure 3.11)

Classic Water Wheel

Introduction

The classic water wheel is elegant, quiet, and very peaceful. It can provide power for a small location. However, water wheels are realistically effective only where the flow of water is reasonably fast. The classic water wheel design is simple, but not the most efficient. There are two basic styles of water wheel: the overshot (figure 3.5a) and the undershot (figure 3.5b). In general, the overshot is more efficient than the undershot.

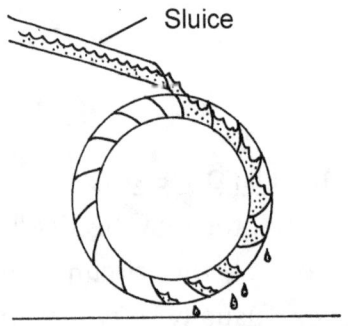

Fig. 3.5a Classic Water Wheel: Fig. 3.5b Classic Water Wheel:
　　　　　Overshot　　　　　　　　　　　　　　　　Undershot

Overshot Water Wheel

The overshot water wheel works by a combination of pressure and gravity. The "blades" of the turbine are actually buckets. The combination of pressure (water hitting the buckets), and the force of gravity (water filling the bucket and weighing it down) moves the buckets downward. The downward motion of the buckets creates a rotation of the wheel.

The efficiency of an overshot water wheel is usually between 60% and 65%. Although the classic water wheel is not efficient enough for large power plants, it can be used for small applications.

The rotational speed is slow, between 6–20 rpm, therefore a series of gears is absolutely required. A smaller wheel can be more effective than a large wheel because wheels with smaller diameter turn faster. Nevertheless, all water wheels require a series of gears in order to get the necessary rotational speed.

Undershot Water Wheel

The undershot wheel is used mostly where the river is shallow. The undershot is far less efficient than the overshot. However, the pressure of the water on the buckets is greater with the undershot. Note that many modern turbines use the principle of water pressure on the buckets (rather than gravity alone) to improve rotational speeds. These modern turbines are in fact advanced forms of the classic undershot.

Pelton Impulse Wheel

Introduction

The Pelton Wheel generally gets higher rotational speeds than the classic water wheel, which means that fewer gears are needed. Often no gears are needed at all. The Pelton is best used on taller dams, at least 150 feet, and operates best with relatively slow flow rates. Under ideal conditions the Pelton Wheel is generally 80% to 90% efficient.

There are three main features of the Pelton Wheel: 1) a high pressure nozzle (the "impulse"), 2) curved buckets, and 3) a ridge in the bucket, splitting the water.

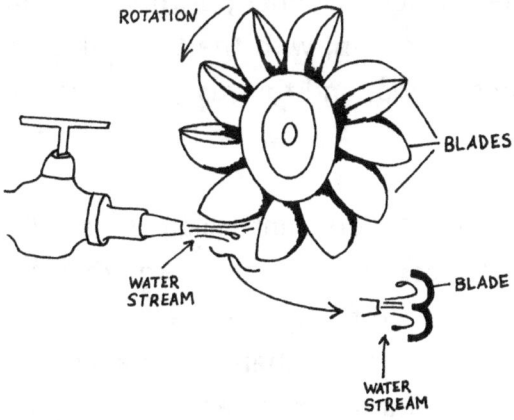

Fig. 3.6 Pelton Impulse Wheel

High Pressure Nozzle: The "Impulse"

The spray nozzle focuses the water into high pressure. The high pressure water then pushes the turbine blades. This is more effective than the classic water wheel. Note that most of the advanced turbine designs have pressure nozzles.

The high pressure nozzle provides greater efficiency than falling water. The classic water wheel works primarily on gravity – the weight of the water on the blades moves the wheel. In contrast, the Pelton Wheel and other advanced designs use pressure of the water to actually push the blades. The pressure from the jet stream is more effective than simple gravity. More jets on the turbine result in a greater amount of water pressure, and therefore a more efficient turbine. Pelton turbines can be built with 1 to 4 jets.

Curved Buckets

In addition, the curved buckets use the force of the water more effectively than just using flat blades of the classic water wheel. Mr. Pelton was not the first inventor to use curved buckets. There is another version of the undershot water wheel, called the Poncelet, which used that design. However, Mr. Pelton made the Poncelet far more effective by combining curved blades with an impulse (jet stream).

A Ridge in the Bucket, Splitting the Water

The Pelton turbine has a ridge in the center of each bucket. In many designs the ridge of the bucket essentially forms *two* buckets. It looks much like a shell that is opened up, with the two halves still connected. This ridge splits the water in two, and the resulting two streams push on their respective buckets in such a way as to make the turbine move faster.

The brief explanation for the efficient result is as follows: Firing the stream on the side of a bucket gives it more energy than firing the stream in the middle of the bucket. Therefore, the ridge exists to make the water hit the two edges, not the center, and therefore makes the turbine rotate faster. (If you wish to understand in greater detail why the ridge makes the Pelton Wheel so effective, you can read other texts.)

Turgo Impulse Wheel

The Turgo is similar to the Pelton but more effective. The efficiencies are similar (80%-90%) but the rotation is twice as fast. As in the Pelton Wheel, the blades of the Turgo Wheel are curved buckets. Also as in the Pelton wheel, a high pressure stream hits the blades of the Turgo. However, the blades of the Turgo are angled differently relative to the high pressure stream. The net result of the design is that the water spray hits 3 blades at once rather than just one.

Fig. 3.7 Turgo Impulse Wheel

Because the jet stream hits 3 blades at once, the turbine rotates faster than if the stream hit just one blade. Advantages of Turgo versus the Pelton are: 1) the Turgo is about half the diameter, 2) the Turgo rotates twice as fast, and 3) the Turgo can be used with dams of lower head (40 – 100 feet).

Propeller Turbine – Fixed Pitch

The propeller turbine is exactly what you'd think. It looks very much like a propeller on a plane or on a boat. The propeller turbine is best used where the flow of water is constant and the head is 30 feet or less.

If you can provide a steady flow of water all the time, then the propeller turbine design is a good choice.

With a steady flow, the efficiency of the propeller turbine is 80% or more. However, if the flow of water is too slow then the propeller becomes very inefficient. If the flow slows down to 30% or less of the design flow, then there will

Fig. 3.8 Propeller Turbine

not be enough force to rotate the turbine at all. In contrast, some other turbines, such as the classic water wheel, the Kaplan, and the Cross–Flow, *can* work with smaller water flows.

Note that *steam* turbines, such as for coal power and nuclear power, operate at a steady flow of steam. Therefore, a propeller design can be very useful in steam turbines.

Kaplan Turbine – Propeller with Adjustable Blade

The Kaplan turbine is actually a form of the propeller turbine. The main difference is that the Kaplan has an adjustable blade. The blades are adjusted to different angles depending on the rate of flow. The adjustable blade allows the turbine to work effectively regardless how slow or how fast the flow of water is. Many Kaplan turbines are automatic: the turbines sense the amount of flow, and adjust the blades accordingly. Kaplan turbines can be installed either horizontally or vertically.

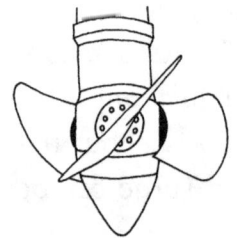

Fig. 3.9 Kaplan Turbine
a) Horizontal Axis

Fig. 3.9 Kaplan Turbine
b) Vertical Axis

Francis Turbine

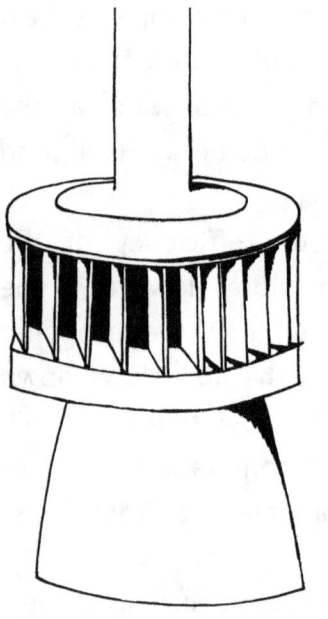

The Francis turbine is the most efficient of turbines. It is the Francis turbine that is used in the large hydropower plants today.

In the Francis turbine, water from the river is directed by a series of adjustable "guide vanes" to the turbine blades.

In general, if we force the water to hit the blades of a turbine at an optimum angle then the water will make the turbine rotate more efficiently. Therefore, the Francis turbine uses adjustable guide vanes to channel water to the turbine blades at the optimum angle at any given time, thus making the turbine rotate very efficiently.

Figure 3.10 Francis Turbine

The Francis turbine also increases the efficiency by using valves to adjust the flow of water flow to an optimum rate. The Francis turbine has a series of valves, known as "wicket gates." These wicket gates control the flow of water. In general, any type of turbine will work most efficiently if the actual water flow is the same as the designed flow. Therefore the wicket gates keep the amount of water flow to the Francis Turbine at ideal values, which ultimately increases the efficiency of the turbine.

The Francis turbine can be used on a dam of any height, and it can work well with the largest flow rates. However, Francis turbines are expensive. The expense of Francis turbines makes them more cost-effective for higher heads than for lower heads. This is because a high head produces more power than a low head, and therefore more electricity will be produced relative to the cost of installation.

Note that the Francis turbine relies on water pressure to turn the blades, which is the same as in the classic water wheel. (This is in contrast to using a high velocity impulse, as was the case in the Pelton.) The Francis turbine is also one of the few turbines which is actually immersed in the water at all times.

The Francis turbine can achieve high efficiencies (80% minimum; often greater) and high rotational speeds (1200 rpm and more). This makes it a very efficient and effective turbine.

Cross-Flow Turbine

The Cross-Flow Turbine is another option where Pelton or Francis turbines are considered. The Cross-Flow Turbine splits the water into two streams. Having two water streams means that the turbine gets two pushes per flow from the reservoir, rather than just one push, which then makes the turbine rotate faster. With multiple streams and the cross-flow of each stream we get multiple pushes of the water on the blades. The net result is that we get more power from each unit of flow coming in from the penstock. This makes the use of water very efficient. The efficiency of the Cross-Flow turbine is 85%-90%.

The Cross-Flow turbine is particularly useful when the actual river flow is less than the design flow. When the water flow is only a portion of the design flow the Cross-Flow can still get high efficiencies. This feature cannot be said of the other turbines.

However, when water flow is faster than the design flow, the Cross-Flow turbine cannot get as high efficiencies as the Francis turbine would. When the flow is very high, the Francis turbine may be the better option.

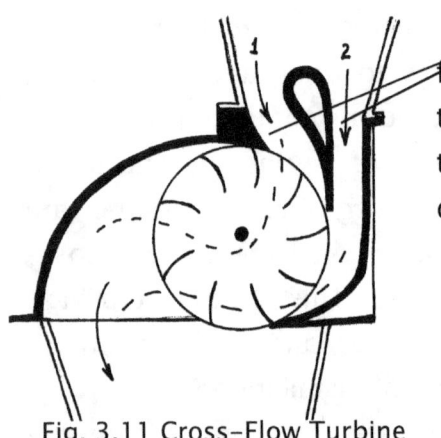

Split Water: The water coming in from the reservoir is split. This results in two streams of water which enter the turbine at different locations and at different times.

Fig. 3.11 Cross-Flow Turbine

General Comparisons of Turbines

Main Points

It is important to understand turbines because using a more efficient turbine will create more electricity. The choice of turbine depends on circumstances such as water flow and head.

Classic Water Wheel

The classic water wheel has low efficiency, around 65%. The classic water wheel needs a head of at least 10 feet. The flow rate can be much lower than all of the other turbines. The upper limits of the water wheel have not been fully tested.

The classic water wheel is generally used in areas where the flow rate is very low, which means that the wheel has a low rotational speed. Higher rotational speeds can be obtained using smaller diameter wheels. However, a series of gears are absolutely required to boost the speed for power generation.

The classic water wheel does have some advantages for small power use. It is quiet, peaceful, and simple to maintain. For small power needs the water wheel can be sufficient.

Pelton

Pelton turbines can work on quite a range of heads, but is best suited for heads over 150 feet. Pelton turbines generally work better with slow flows than with fast flows. Pelton turbines have a good efficiency of 80% to 90%.

Turgo

The Turgo turbine can work with quite a range of heads and range of flow rates. The usefulness of the Turgo turbine is similar to the Pelton in most respects, with a few distinctions. The Turgo turbine rotates faster than the Pelton, usually requiring no gears. The Turgo has a better efficiency with changing flows than the Pelton. Specifically, if the flow is slower than the design flow value, the Turgo is more efficient.

Propeller

Propeller turbines work best where there is a low head (less than 30 feet) and a constant flow. Constant flow is the real key to using a propeller turbine. If you can control the flow then a propeller turbine works well.

Kaplan

The Kaplan turbine works best with heads less than 250 feet. The Kaplan can work with a wide range of flow speeds, second only to the range of speeds offered by the Francis turbine. One primary advantage of the Kaplan turbine over other turbines is that the Kaplan can have a reasonable efficiency regardless of the changing flow.

Francis

In general, Francis turbines are used on high heads and fast flows. Francis turbines work on quite a range of heads, however the expense of Francis turbines makes them more cost-effective for higher heads than for lower heads. Francis turbines are the best choice (and have the best efficiency) for fast flows. Francis turbines are capable of taking the fastest flows anywhere. The Francis Turbine is the ideal choice for megawatt hydropower plants.

Cross-Flow

The Cross-Flow turbine can work on a wide range of heads, but is most effective with heads of 250 feet or lower. The Cross-Flow turbine generally works best at lower flow rates than the other turbines. However, the Cross-Flow has an advantage over the other turbines in that it has a higher efficiency with partial flow than the other turbines.

Turbines Compared: Data Comparisons

Introduction

The data below looks at each type of turbine, citing the range of head values and the range of flow values where that type of turbine is most effective.

Each table provides the same data, but using different units. Providing the same data in different units is necessary because reports might use any of these units.

Also note that some of the larger numbers have been rounded to a value ending in zero. Rounding to a zero for larger values in this table can be acceptable. However, smaller values will not be rounded in the same way because such rounding could make errors on a practical level.

A. Best ranges for turbines: head in meters, flow in m³/s

Turbine	Head (in meters)	Flow (in m³/s)
Kaplan	2.13 m – 79.3 m	.001 – 50 m³/s
Cross-Flow	1.2 m – 201 m	.047 – 12.8 m³/s
Francis	3 m – 701 m	.5 –1,000 m³/s
Turgo	12.2 m – 275 m	.001 – 10 m³/s
Pelton	50 m – 1000 m	.001 – 2 m³/s

B. Best ranges for turbines: head in meters, flow in Liters/sec

Turbine	Head (in meters)	Flow (in Liters/sec)
Kaplan	2.13 m – 79.3 m	1 – 50,000 L/s
Cross-Flow	1.2 m – 201 m	47.2 – 12,860 L/s
Francis	3 m – 701 m	500 – 1,000,000 L/s
Turgo	12.2 m – 275 m	1 – 10,000 L/s
Pelton	50 m – 1000 m	1 – 2,000 L/s

C. Best ranges for turbines: head in feet, flow in ft^3/s ("cfs")

Turbine	Head (in feet)	Flow (in ft^3/sec)
Kaplan	7 ft – 260 ft	.035 – 1,750 ft^3/sec
Cross–Flow	4 ft – 660 ft	1.66 – 450 ft^3/sec
Francis	10 ft – 2,300 ft	17.5 – 35,000 ft^3/sec
Turgo	40 ft – 902 ft	.035 – 350 ft^3/sec
Pelton	164 ft – 3,280 ft	.035 – 70 ft^3/sec

D. Best ranges for turbines: head in feet, flow in ft^3/min ("cfm")

Turbine	Head (in feet)	Flow (in ft^3/min)
Kaplan	7 ft – 260 ft	2.1 –105,000 ft^3/min
Cross–Flow	4 ft – 660 ft	100 – 27,000 ft^3/min
Francis	10 ft – 2,300 ft	1,050 – 2,100,000 ft^3/min
Turgo	40 ft – 902 ft	2.1 –21,000 ft^3/min
Pelton	164 ft – 3,280 ft	2.1 – 4,200 ft^3/min

E. Best ranges for turbines: head in feet, flow in gallons/min ("gpm")

Turbine	Head (in feet)	Flow (in gal/min)
Kaplan	7 ft – 260 ft	15.8 – 793,000 gal/min
Cross–Flow	4 ft – 660 ft	750 – 204,000 gal/min
Francis	10 ft – 2,300 ft	8,000 – 15,875,000 gal/min
Turgo	40 ft – 902 ft	15.8 – 158,700 gal/min
Pelton	164 ft – 3,280 ft	15.8 – 31,750 gal/min

Conclusion

Many Americans hold passionate views about electrical power, yet few Americans understand all the details behind their passion. Electricity should not be mysterious. The science, the technology, and the data of electrical power can be understood by anyone.

Above all else, we must remember that there are no perfect solutions, there are only choices. Any option can be beneficial, yet each option has its own technical issues to work with. It is up to you and to your community to make those educated decisions. I hope that this book will help guide you in your choices.

M.F.

Flow Unit Equivalents

A. <u>1 gallon/min is equal to</u>:

1 gal/min   = 1.0 gal/min = .133 ft^3/min   = .063 kg/s = .063 L/s
           = .002 ft^3/s = .000062 m^3/s

B. <u>1 ft^3/min is equal to</u>:

1 ft^3/min   = 7.48 gal/min = 1.0 ft^3/min = .471 kg/s   = .471 L/s = .016 ft^3/s
           = .000471 m^3/s

C. <u>1 kg/s is equal to</u>:

1 kg/s   = 15.9 gal/min = 2.12 ft^3/min = 1.0 kg/sec = 1.0 L/sec
        = .035 ft^3/s = .001 m^3/s

D. <u>1 L/s is equal to</u>:

1 L/s   = 15.9 gal/min = 2.12 ft^3/min = 1.0 kg/sec = 1.0 L/s
       = .035 ft^3/s   = .001 m^3/s

E. <u>1 ft^3/s is equal to</u>:

1 ft^3/s   = 448.8 gal/min   = 60 ft^3/min = 28.5 kg/s = 28.5 L/s
         = 1.0 ft^3/s = .0285 m^3/s

F. <u>1 m^3/s is equal to</u>:

1 m^3/s   = 15,840 gal/min   = 2,118 ft^3/min = 1,000 kg/s = 1,000 L/s
         = 35.3 ft^3/s = 1.0 m^3/s

Flow Conversions
From Any Unit into Liters/sec

A. from gallon/min to Liters/sec: # Liters/sec = .063 x # gallons/min

B. from ft^3/min to Liters/sec: # Liters/sec = .47 x # ft^3/min

C. from kg/s to Liters/sec: # Liters/sec = 1.0 x # kg/s

D. from ft^3/s to Liters/sec: # Liters/sec = 28.5 x # ft^3/s

E. from m^3/s to Liters/sec: # Liters/sec = 1,000 # m^3/s

Bibliography

Hydroelectric Power

1. Energy for Man: From Windmills to Nuclear Power, by Hans Thirring, 1958. Publisher: Indiana University Press.
2. Energy Resources, by Andrew Simon, 1975. Pergamon Press, Inc.
3. Nontechnical Guide to Energy Resources, by Ben Ebenhack, 1995. Publisher: PennWell Publishing Company
4. Electric Power Generation: A Nontechnical Guide, by Barnett and Bjornsgaard, 2000. Publisher: PennWell Publishing Company
5. Energy: A Guidebook, by Janet Ramage, 1997. Oxford University Press.
6. Wind and Water Power, by Clint Twist, 1993. Publisher: Gloucester Press
7. Energy: A First Reference Book, by Melvin Berger, 1983. Franklin Watts
8. Harnessing Water Power for Home Energy, by Dermot McGuigan, 1978. Publisher: Garden Way Publishing Co.
9. microhydropower.net www.microhydropower.net
10. Layman's Guidebook on how to develop a small hydro site, European Small Hydropower Association (ESHA), http://europa.eu.int/comm/energy/library/hydro/layman2.pdf
11. Energy Educators of Ontario http://www.iclei.org/efacts/hydroele.htm
12. Association of State Dam Safety Officials http://www.damsafety.org/
13. Lake Sedimentation Project http://www3.baylor.edu/Geology/lake/lake.html
14. National Sedimentation Laboratory (in USDA) http://msa.ars.usda.gov/ms/oxford/nsl/
15. Canyon Industries www.canyonindustriesinc.com/
16. Database of Dams, US Bureau of Reclamation http://www.usbr.gov/dataweb/dams/
17. National Inventory of Dams, US Army Corp of Engineers, http://crunch.tec.army.mil/nid/webpages/nid.cfm
18. List of Hydropower Plants, US Bureau of Reclamation http://www.usbr.gov/power/facil/facil.html
19. Hoover Dam, US Bureau of Reclamation http://www.usbr.gov/lc/hooverdam
20. Grand Coulee Dam, US Bureau of Reclamation www.usbr.gov/power/data/sites/grandcou/grandcou.html
21. Waterwheel Factory www.waterwheelfactory.com/index.htm
22. Canyon Industries www.canyonindustriesinc.com/

Government Sites – General

1. US Department of Energy (DOE) www.energy.gov
2. US Department of the Interior www.doi.gov
3. US Bureau of Reclamation www.usbr.gov
4. US Department of Agriculture (USDA) www.usda.gov
5. Environmental Protection Agency (EPA) www.epa.gov
6. Food and Drug Administration (FDA) www.cfsan.fda.gov
7. National Institute for Occupational Safety and Health (NIOSH)
   www.cdc.gov/niosh
8. Mine Safety and Health Administration (MSHA) www.msha.gov
9. Federal Energy Regulatory Commission (FERC) www.ferc.gov
10. Nuclear Regulatory Commission (NRC) www.nrc.gov
11. National Climatic Data Center (NCDC) www.ncdc.noaa.gov

Department of Energy (DOE) Related Sites

1. Department of Energy (DOE) www.energy.gov
2. Energy Information Administration (EIA) www.eia.doe.gov
3. [Office of] Efficiency and Renewable Energy (EERE) www.eere.energy.gov
4. Office of Fossil Energy (in Dept of Energy) www.fossil.energy.gov
5. Electric Transmission and Distribution Office www.electricity.doe.gov
6. Science (Office of Science) www.sc.doe.gov
7. Nuclear Regulatory Commission (NRC) www.nrc.gov
8. Civilian Radioactive Waste Management (OCRWM) www.ocrwm.doe.gov
9. Yucca Mountain Project www.ocrwm.doe.gov/ymp/about/index.shtml
10. International Nuclear Safety Program http://insp.pnl.gov
11. International Nuclear Safety Center, Argonne Laboratory www.insc.anl.gov
12. National Energy Technology Laboratory (NETL) www.netl.doe.gov
13. National Renewable Energy Laboratory (NREL) www.nrel.gov
14. Oak Ridge National Laboratory www.ornl.gov
15. Los Alamos National Laboratory (LANL) www.lanl.gov/worldview
16. Pacific Northwest National Laboratory (PNL) www.pnl.gov
17. Starlight, from PNNL/DOE http://starlight.pnl.gov

Index

www.ingramcontent.com/pod-product-compliance
Lightning Source LLC
Chambersburg PA
CBHW081357170526
45166CB00010B/3112